Look into the HUMAN BODY

RESPIRATORY SYSTEM

TRACY VONDER BRINK

A Crabtree Crown Book

Crabtree Publishing
crabtreebooks.com

School-to-Home Support for Caregivers and Teachers

This appealing book is designed to teach students about core subject areas. Students will build upon what they already know about the subject, and engage in topics that they want to learn more about. Here are a few guiding questions to help readers build their comprehension skills. Possible answers appear here in red

Before Reading:

What do I know about the respiratory system?

- *I know the respiratory system is part of my body.*
- *I know my lungs are part of my respiratory system.*

What do I want to learn about this topic?

- *I want to know how the respiratory system works.*
- *I want to learn about the parts of the respiratory system.*

During Reading:

I'm curious to know...

- *I'm curious to know how my body uses oxygen.*
- *I'm curious to know why I breathe out carbon dioxide.*

How is this like something I already know?

- *I know I breathe in oxygen.*
- *I know I inhale and exhale.*

After Reading:

What was the author trying to teach me?

- *The author was trying to teach me what my respiratory system does.*
- *The author was trying to teach me how my lungs work.*

How did the images and captions help me understand more?

- *The images helped me understand how the parts of the respiratory system work together.*
- *The captions gave me extra information about each part of the respiratory system.*

TABLE OF CONTENTS

CHAPTER 1

MEET YOUR BODY

Have you ever wondered how your body takes in air, moves your blood, digests your lunch, and reads this book...all at the same time? It begins with cells.

A cell is the basic unit of life. You are made up of trillions of cells! Different cells have different kinds of jobs. For example, alveolar cells in your lungs take in **oxygen**. The human body has around 200 different kinds of cells.

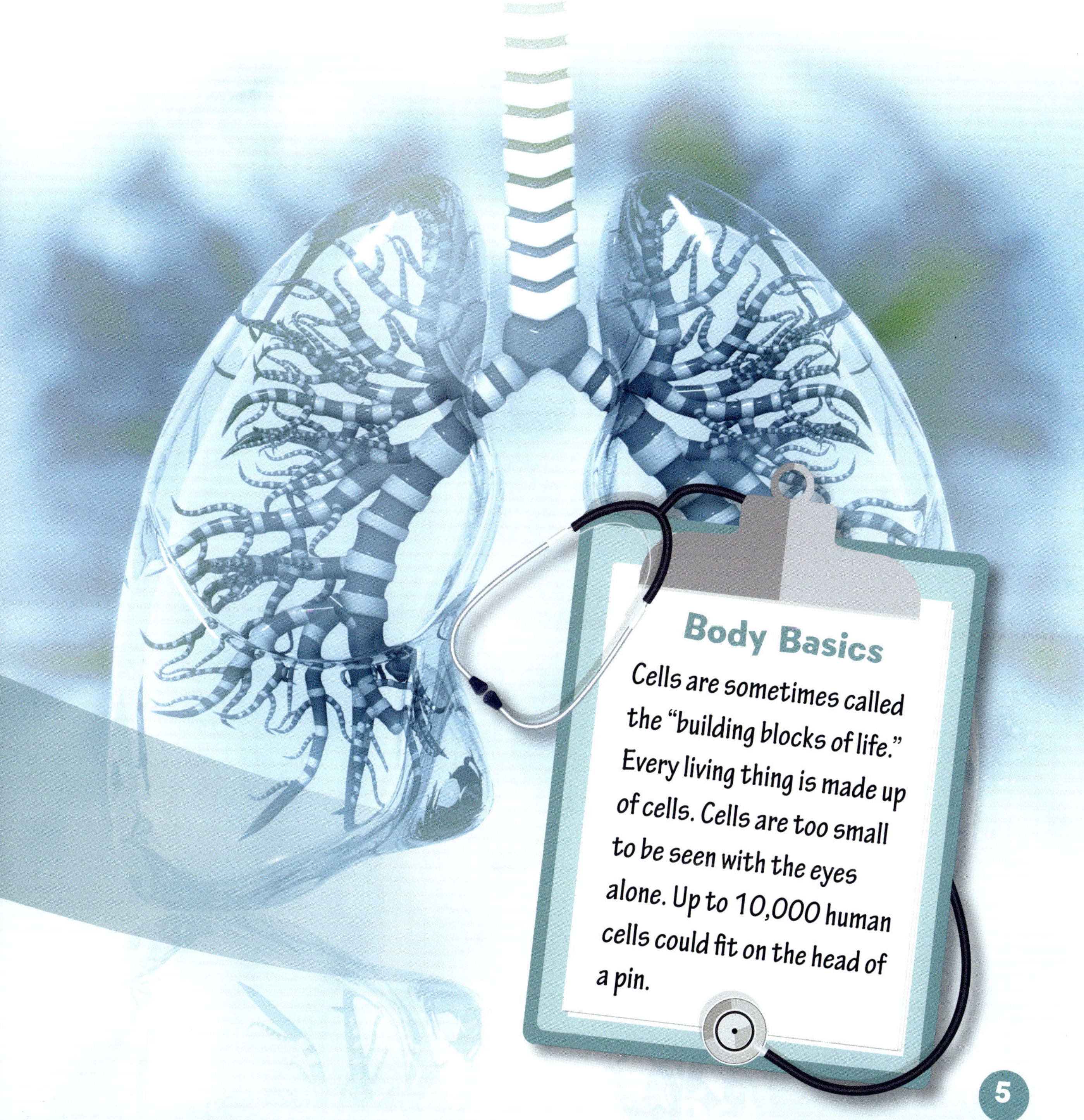

Body Basics

Cells are sometimes called the "building blocks of life." Every living thing is made up of cells. Cells are too small to be seen with the eyes alone. Up to 10,000 human cells could fit on the head of a pin.

Your body has many different **structures** made of cells. Your lungs are one of these structures. Different structures make up body systems. Each system has important tasks. The structures in a system work together to keep your body running.

CHAPTER 2

YOUR RESPIRATORY SYSTEM

The respiratory system is a network of **organs** and **tissues** that helps you breathe. The parts of the respiratory system work together so you can move air in and out of your body. Your respiratory system is working all the time.

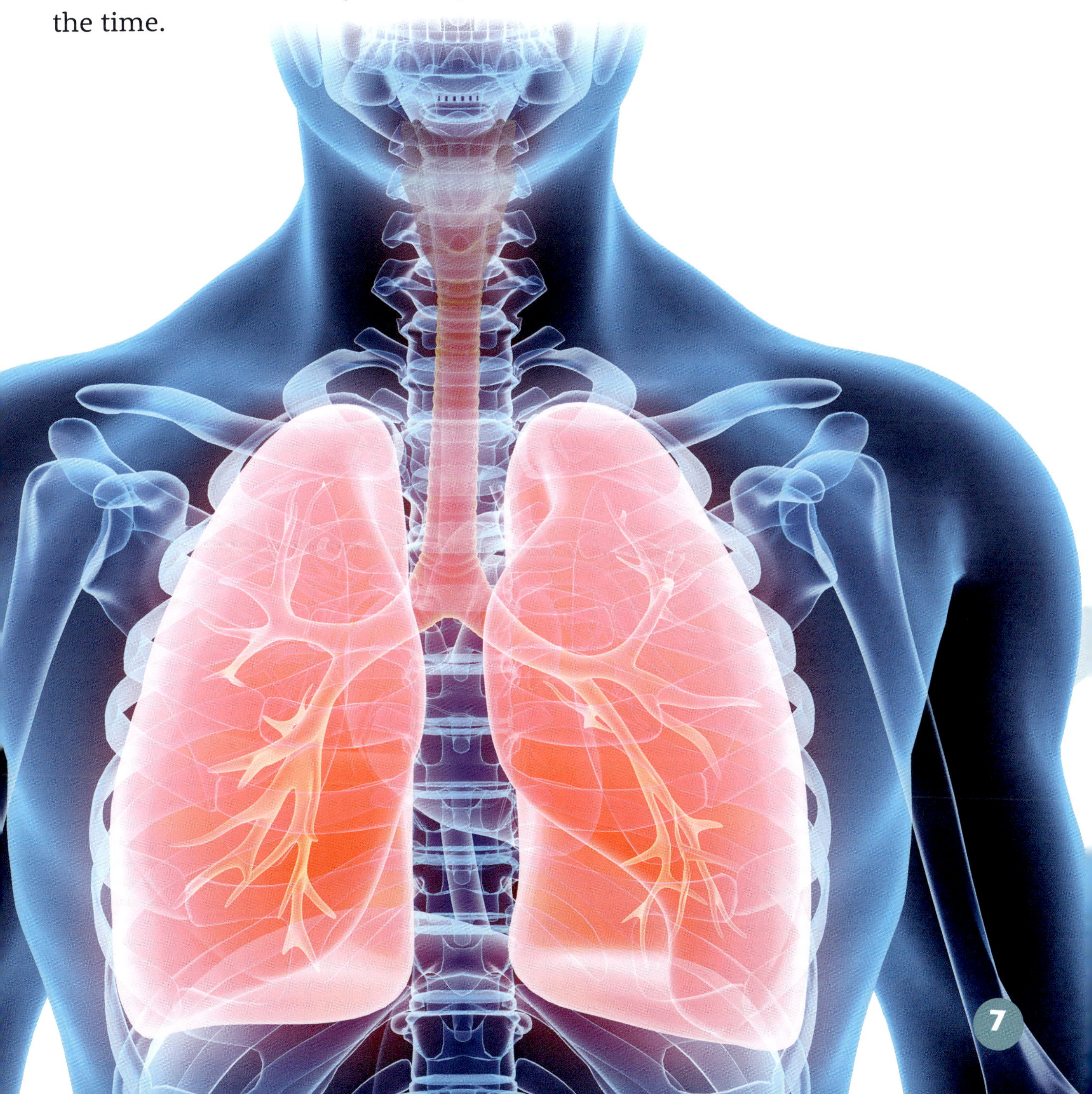

Cells must have oxygen to live. Your respiratory system brings in the oxygen they need. As cells work, they make waste gases such as **carbon dioxide**. Your respiratory system helps your body clean out this waste.

You inhale oxygen with every breath. Then you exhale carbon dioxide.

When you smell something, your nose and brain work together to identify the scent.

Your respiratory system also allows you to talk and smell. You talk and smell when air moves through parts of your respiratory system.

CHAPTER 3

PARTS OF YOUR RESPIRATORY SYSTEM

Body Basics

Smells pass into your nose with the air you breathe. Cells in your nose send the information to your brain so you notice the smell. Your nose can detect around 1 trillion different smells!

YOUR NOSE

Your nose is an opening where you breathe air in and out. A sticky **membrane** covers the inside of your nose. This membrane is also covered in thousands of tiny hairs. Together, they trap dust and dirt to stop them from entering your lungs.

YOUR SINUSES

Your sinuses are hollow spaces inside your skull. They are filled with air. The sinuses make the **mucus** that keeps your nose moist. Your sinuses also **filter** the air you breathe.

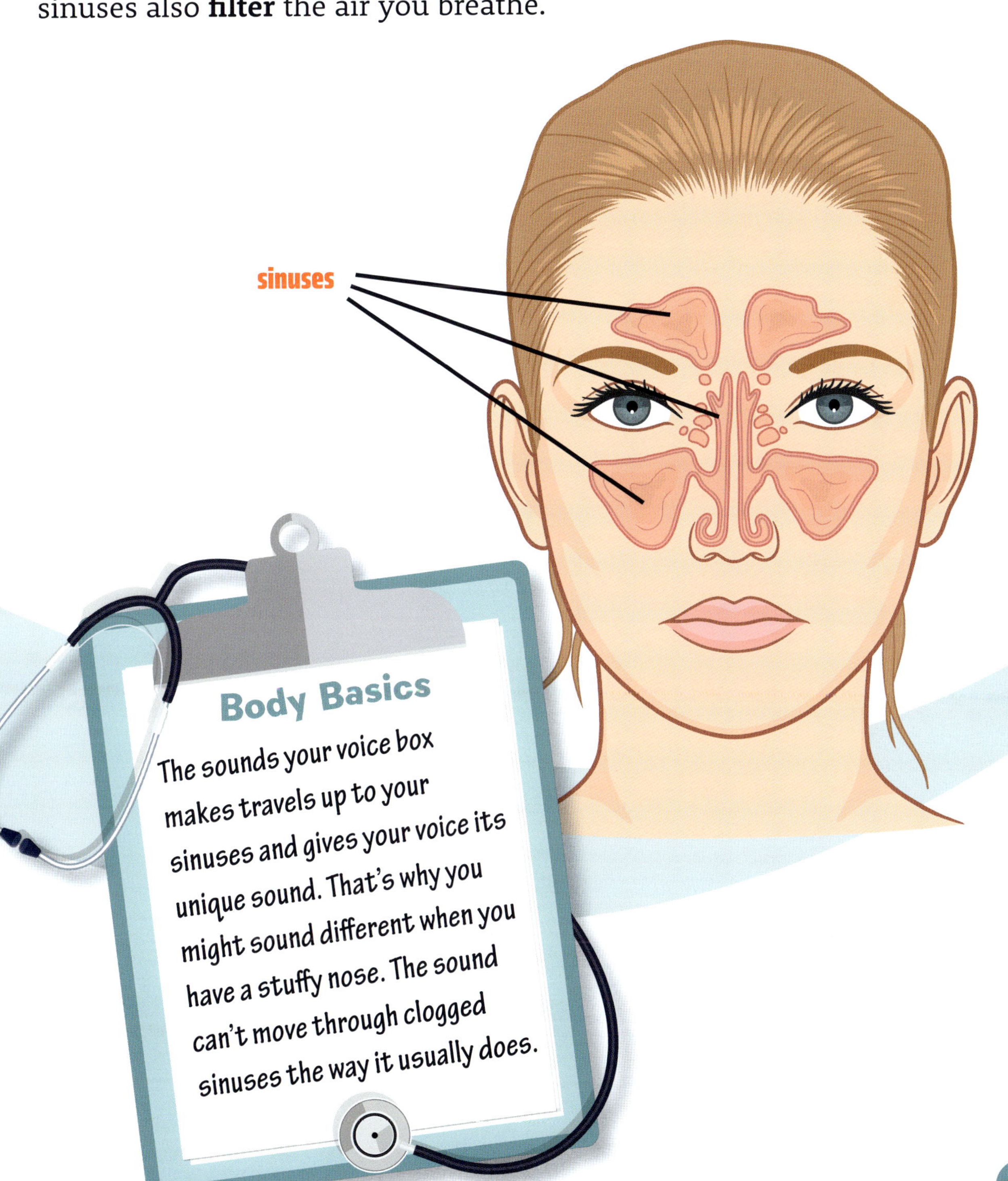

YOUR PHARYNX

The pharynx is a tube that starts behind the nose. It carries air to your trachea, or windpipe. It also carries food and drink from your mouth to your esophagus, which connects to your stomach.

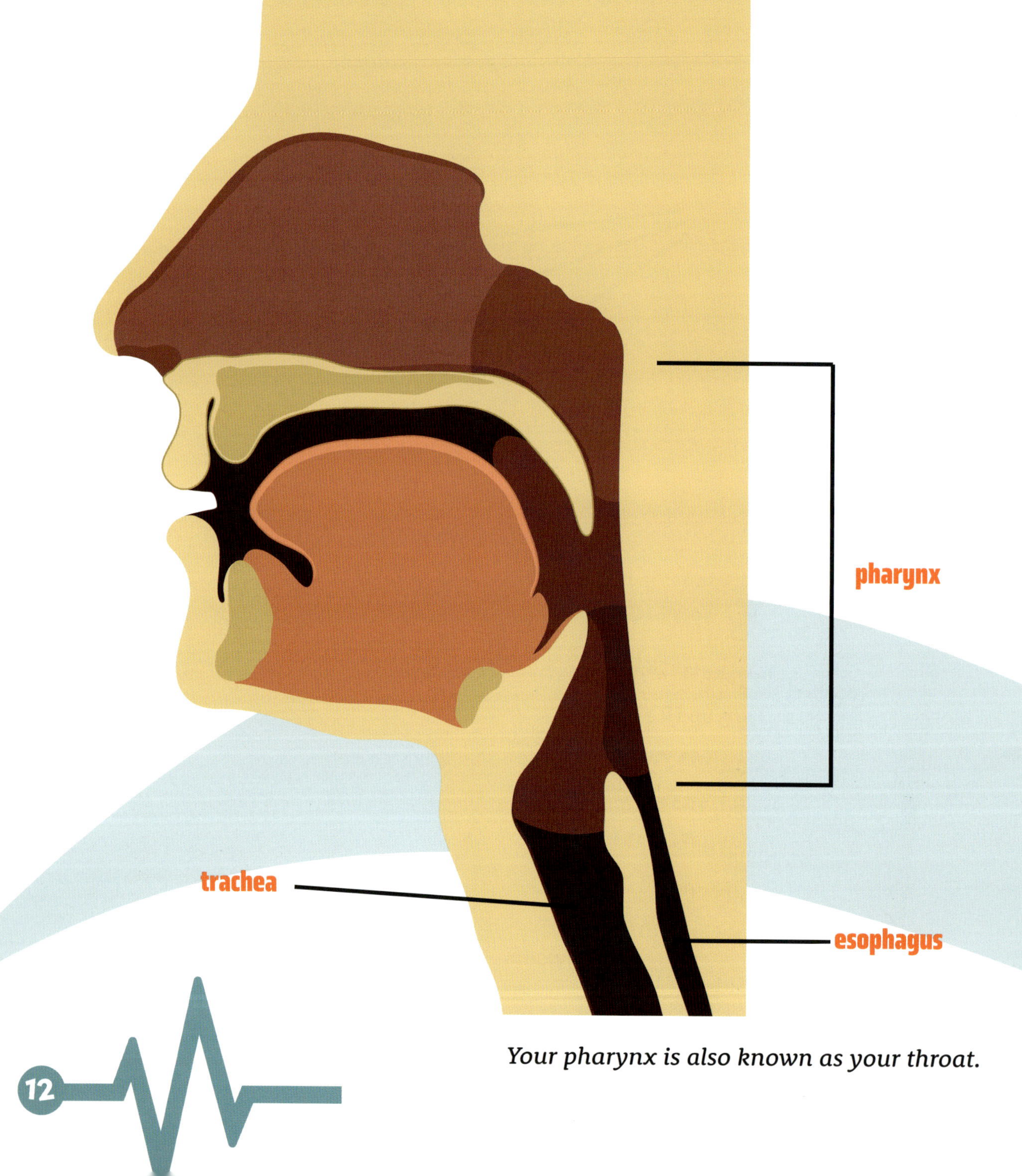

Your pharynx is also known as your throat.

YOUR TRACHEA

The trachea is a passage that connects your throat and lungs. It branches into two tubes. One tube connects to each lung. A flap called the epiglottis folds over the trachea to stop food and drink from entering the lungs.

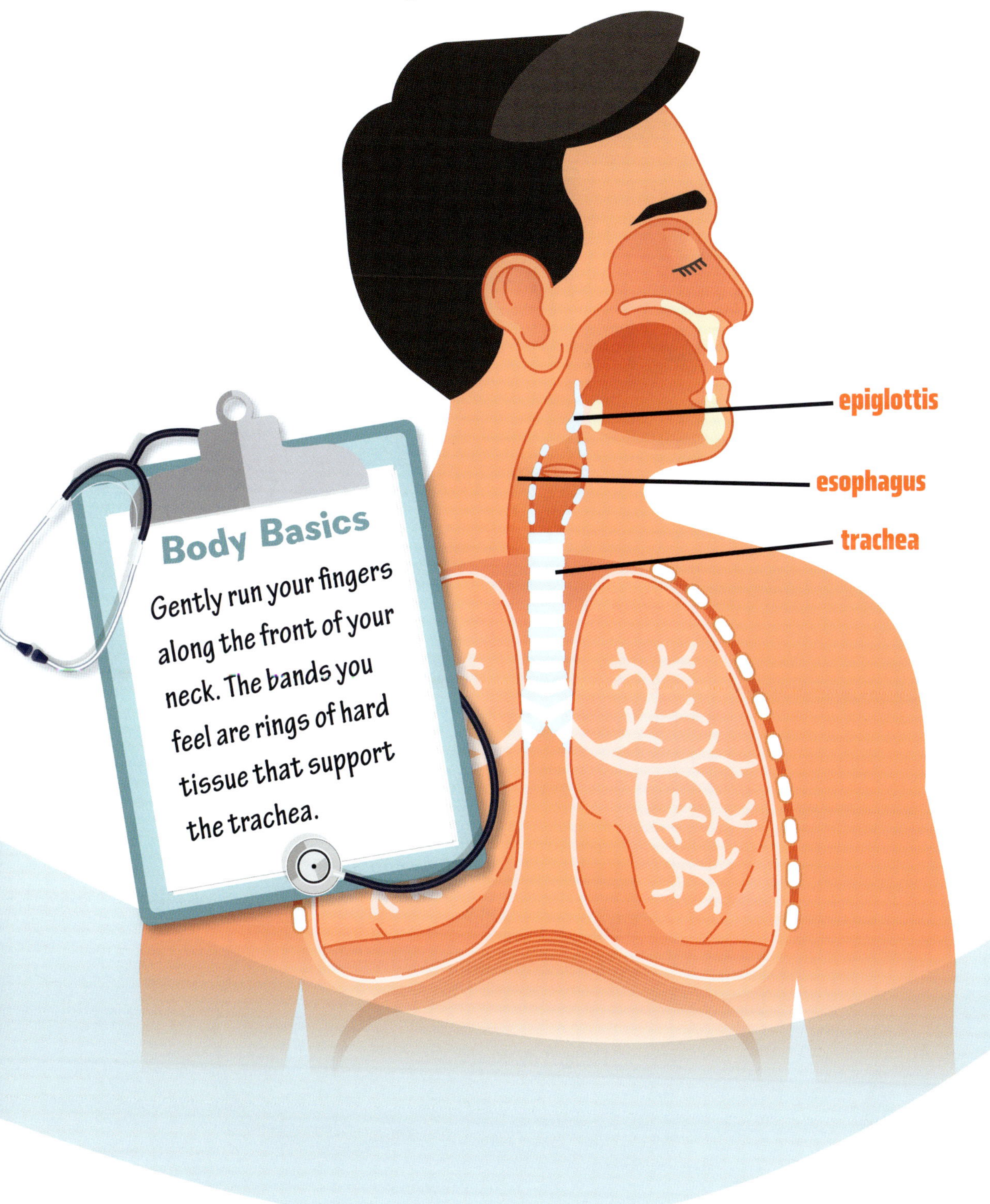

YOUR LARYNX

Your larynx is at the top of the trachea. It's also called the voice box. The larynx holds the vocal folds. These folds **vibrate** when air passes through them. The vibrations are how you talk, hum, and sing.

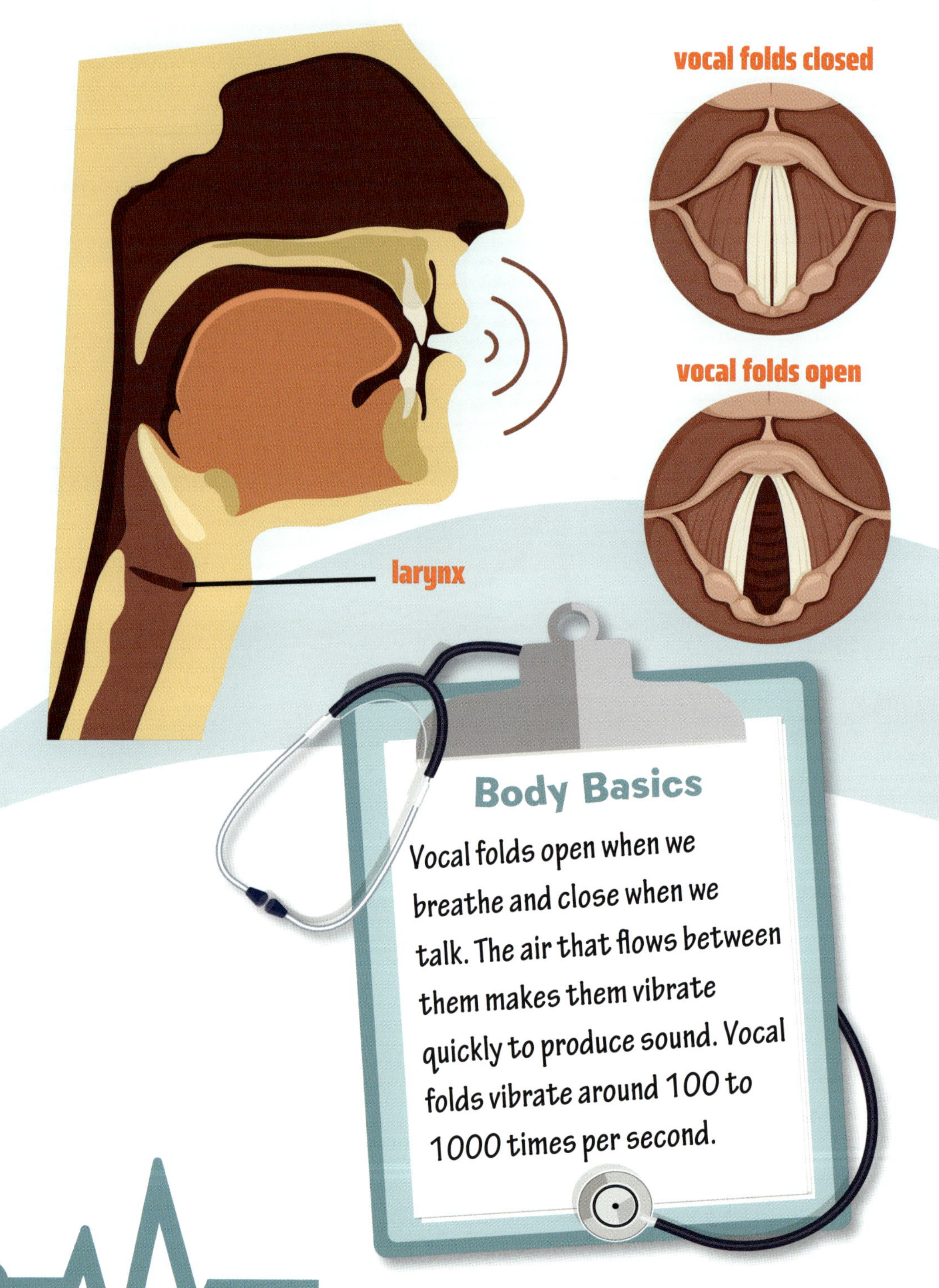

YOUR LUNGS

The lungs are made of tissue that stretches so you can breathe in and out. A muscle below the lungs called the diaphragm helps them pull in and push out air. Many airways branch through both lungs. These airways are filled with millions of tiny air sacs.

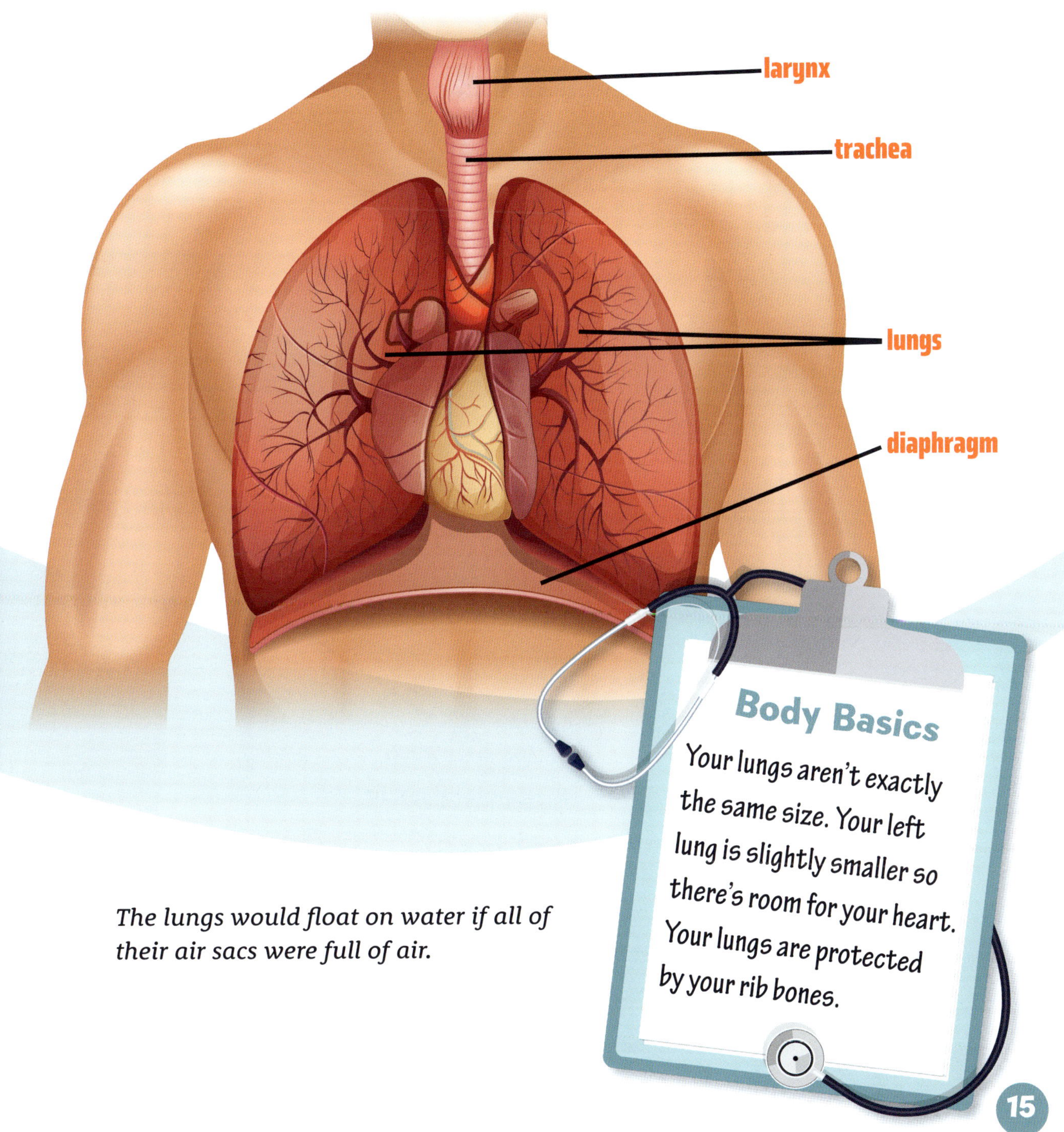

The lungs would float on water if all of their air sacs were full of air.

Body Basics

Your lungs aren't exactly the same size. Your left lung is slightly smaller so there's room for your heart. Your lungs are protected by your rib bones.

CHAPTER 4:

THE RESPIRATORY SYSTEM AT WORK

Breathing in starts with the diaphragm. When you breathe in, your diaphragm moves and expands your ribcage. This makes more room in your chest to pull air through the mouth and nose to your lungs.

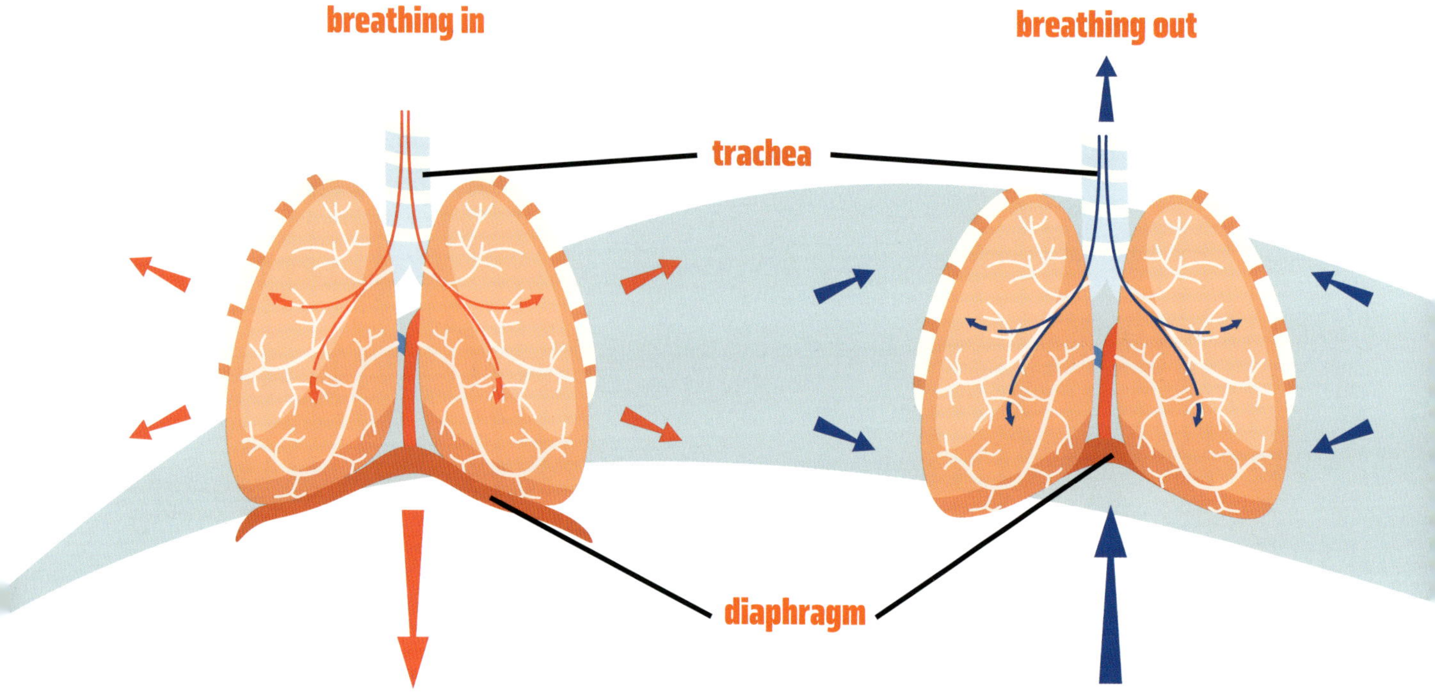

breathing in

breathing out

trachea

diaphragm

Oxygen moves through the trachea and into the lungs as you breathe in. Carbon dioxide leaves the lungs and moves through the trachea when you breathe out.

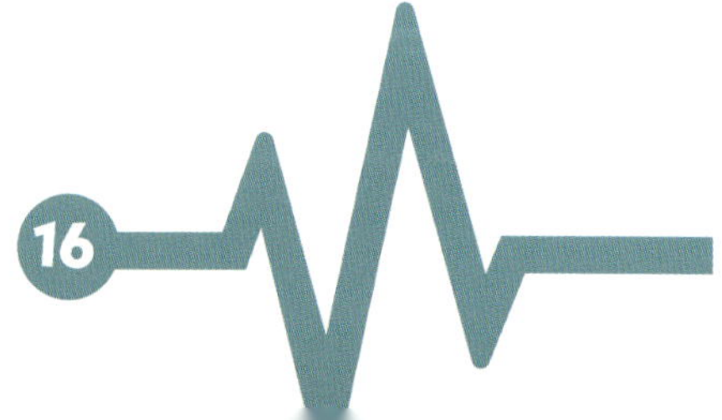

Oxygen fills your air sacs with every breath. It moves from the air sacs and into your blood. The blood flows to your heart, and the heart pumps it out into the rest of the body. The oxygen in the blood goes to the cells.

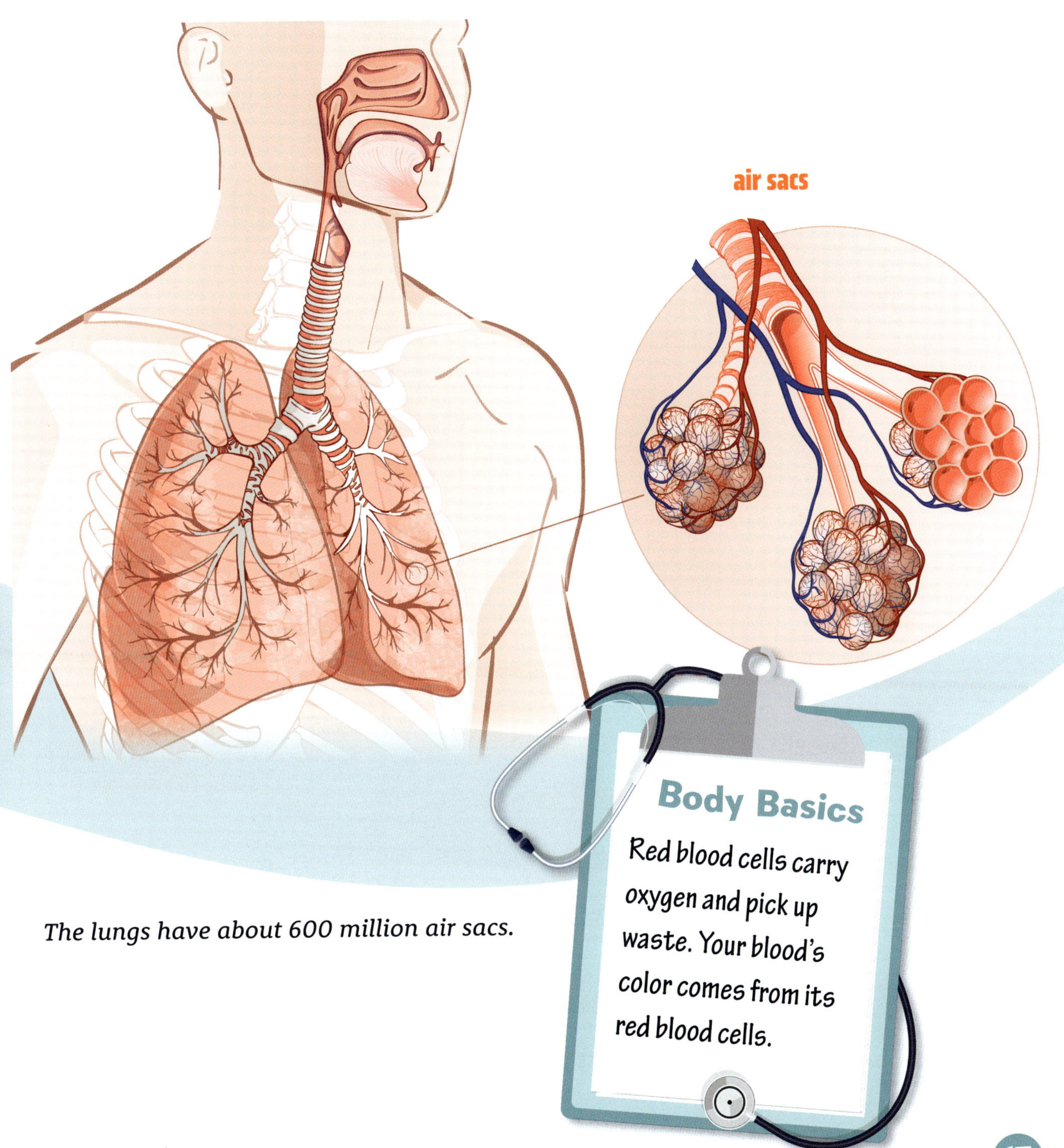

The lungs have about 600 million air sacs.

Body Basics

Red blood cells carry oxygen and pick up waste. Your blood's color comes from its red blood cells.

Carbon dioxide and other waste gases pass from your body's cells and into your blood. It flows back to the heart. Then, the heart pumps it into the lungs. In the lungs, the carbon dioxide goes into the air sacs. It leaves your body when you breathe out.

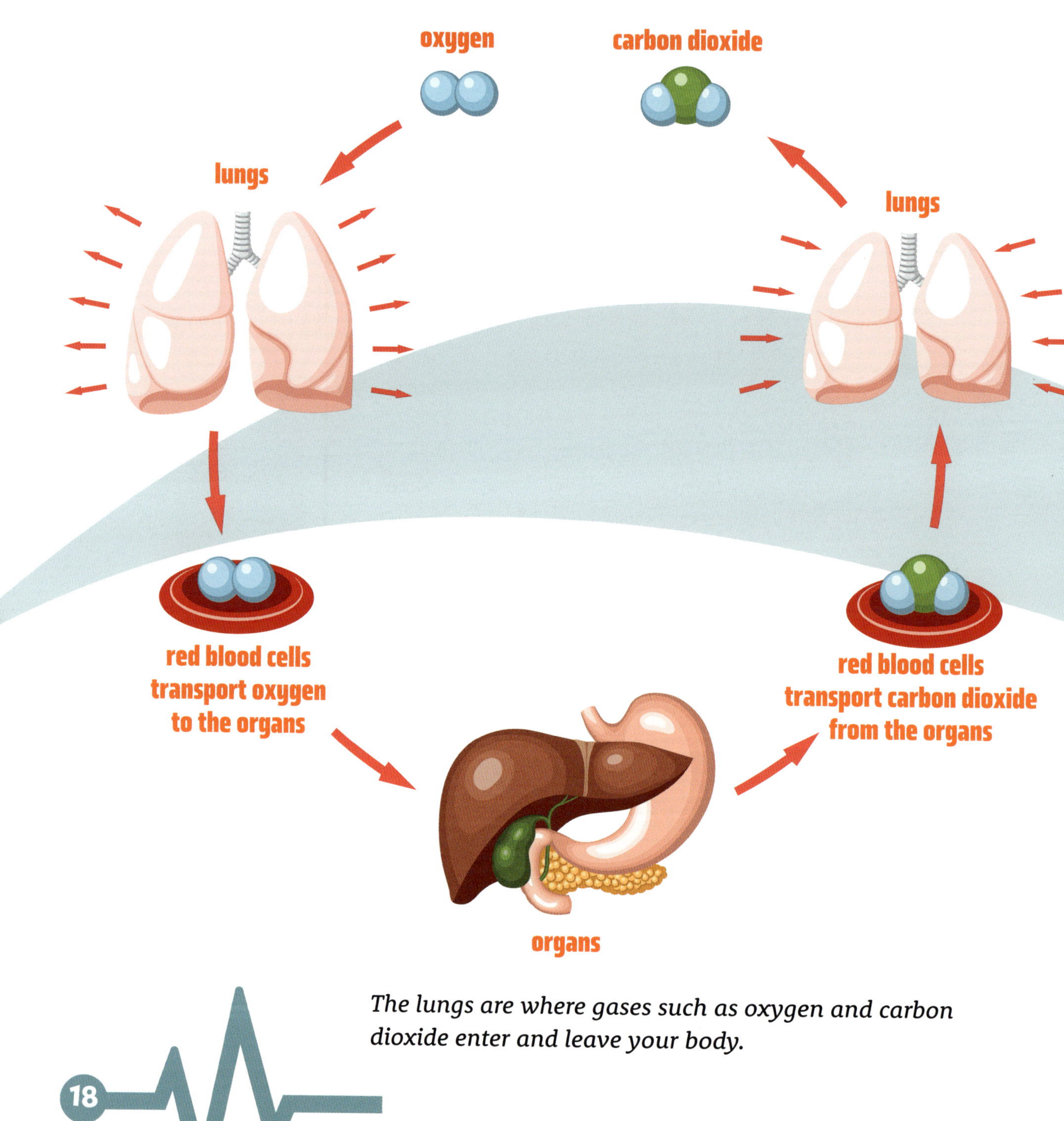

The lungs are where gases such as oxygen and carbon dioxide enter and leave your body.

This **cycle** happens all the time, day and night. You breathe in and out about 22,000 times every day! Breathing speeds up if your body needs more oxygen, such as when you run a race. When you're resting and need less oxygen, your breathing slows down.

Body Basics

Lungs can take in oxygen from air, but they can't take it from water. That's why people can't breathe underwater. Fish have gills that can take oxygen from water, but they can't breathe air.

CHAPTER 5

RESPIRATORY SYSTEM PROBLEMS

ALLERGIES

Your body's **immune** system fights germs that might make you sick. But sometimes it mistakes harmless dust or **pollen** for these germs. Then it signals the nose and lungs to make mucus and flush them out. That's why people with allergies cough and have runny noses, even though they aren't sick.

ASTHMA

Asthma is a condition that causes the airways in the lungs to swell. The swollen airways make it harder to breathe. A person with asthma might wheeze when they breathe. Sometimes, asthma is triggered by allergies or illness. People with asthma might also wheeze when they exercise.

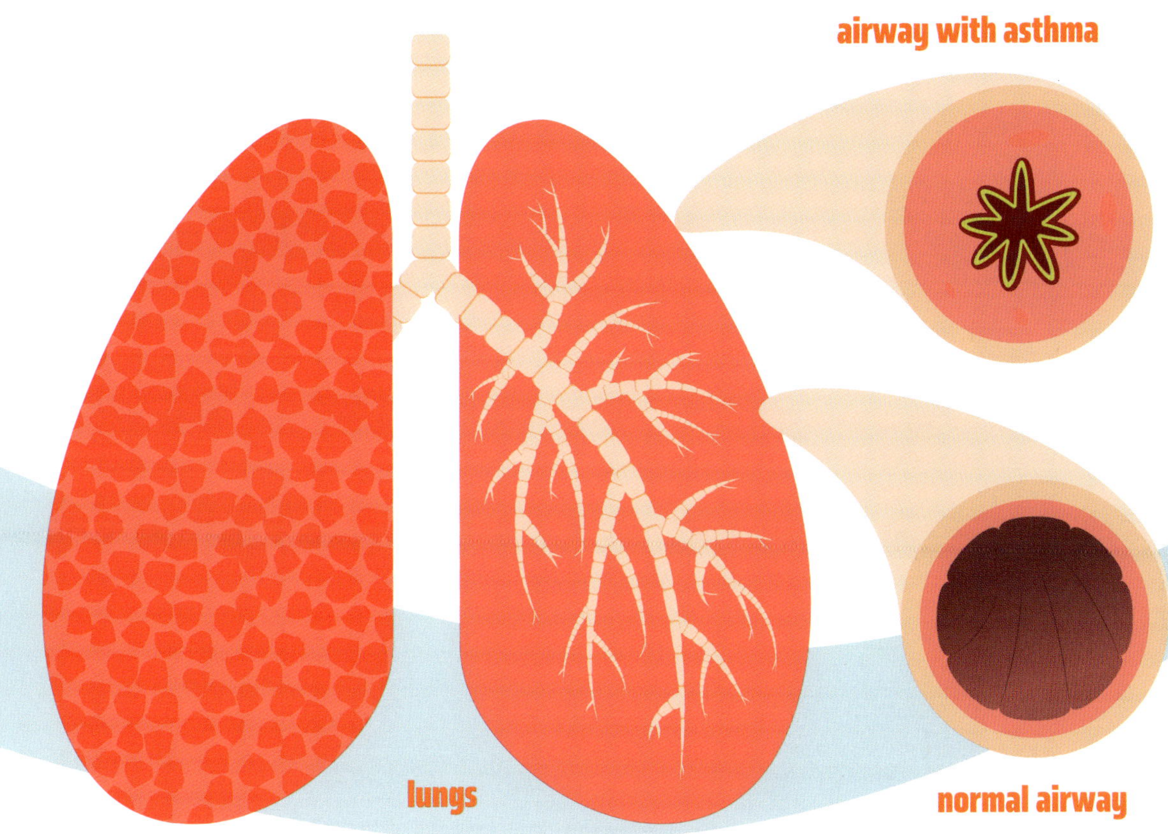

RESPIRATORY INFECTIONS

Some germs attack the respiratory system. They cause sore throats, coughs, and colds. Pneumonia is a serious infection of the lungs. COVID-19 is also a respiratory infection.

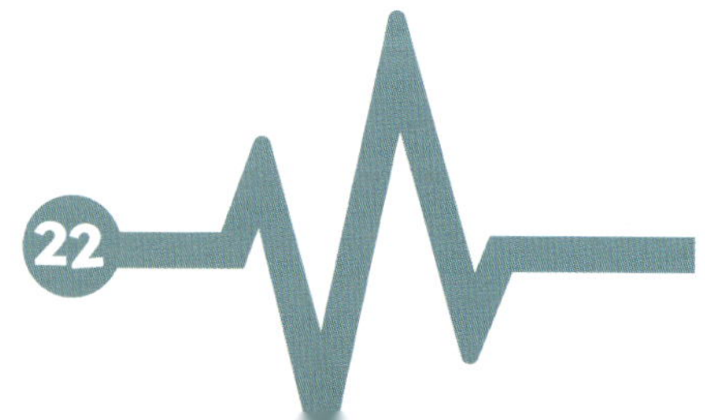

Body Basics

Asthma can't be cured, but it can be controlled with medicine. Many people with asthma use an inhaler, a device that lets them breathe in medicine to help their lungs.

HOW DOCTORS HELP

Doctors can look in your nose and throat and listen to your lungs to find any problems with your respiratory system. They may use a chest X-ray to check for pneumonia. Some respiratory problems go away on their own. Others need medicine.

CHAPTER 6

TAKING CARE OF YOUR RESPIRATORY SYSTEM

BE ACTIVE

Being active makes your lungs work harder and become stronger. Activities such as jumping rope, riding a bike, swimming, and running all strengthen the lungs. Even walking quickly helps.

EAT AND DRINK WELL

Eating healthy foods helps your whole body, including your respiratory system. Drinking enough water helps thin the mucus your body makes. Thinner mucus makes it easier for your lungs to work.

NEVER SMOKE

Smoking hurts the lungs. It destroys lung tissue. People who smoke have a harder time breathing. Smoking is also the major cause of lung **cancer**.

LUNG CANCER

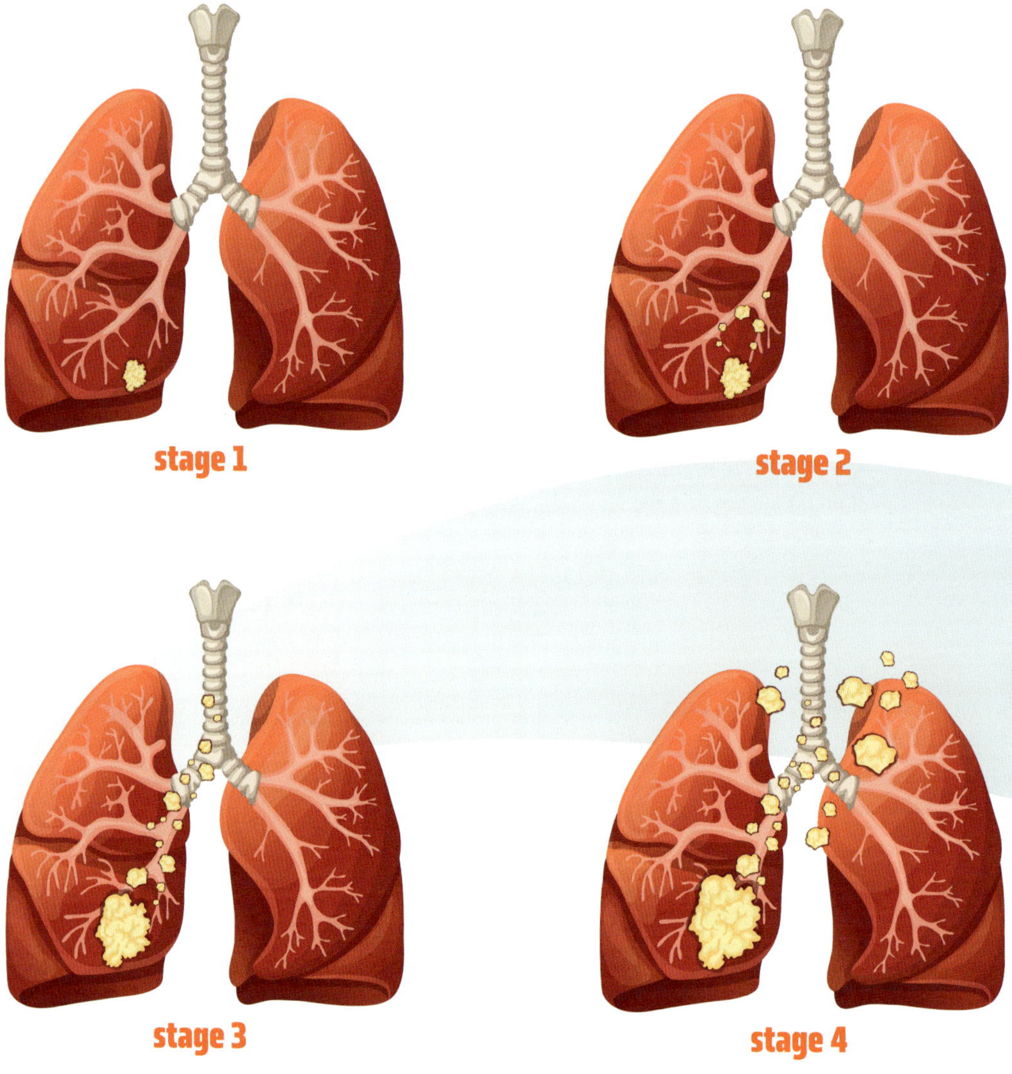

stage 1

stage 2

stage 3

stage 4

Cancer is a disease in which abnormal cells grow out of control. Cancer cells can spread to other parts of the body.

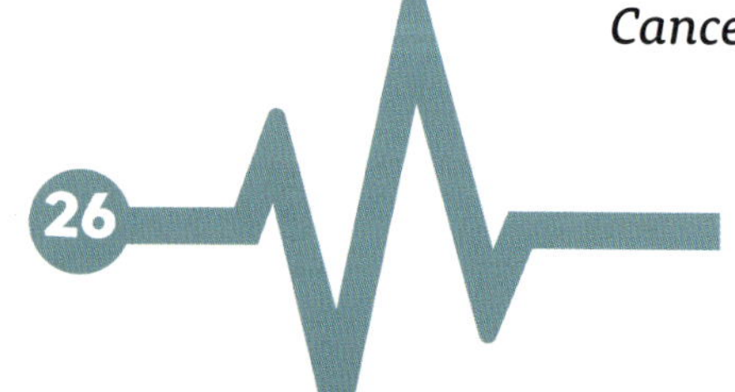

SEE YOUR DOCTOR

Regular check-ups help keep you healthy. The doctor listens to your lungs and looks in your mouth and nose to make sure everything is working right. They will help you if you are sick or have problems breathing.

Body Basics

Germs spread when a sick person sneezes or coughs them out, and you breathe them in. If you touch somewhere germs land, the germs stick to your fingers. That's why you shouldn't touch your eyes, mouth, or nose unless your hands are clean.

Help stop germs by washing your hands with soap and water, or use hand sanitizer. Try not to touch your mouth, nose, or eyes if your hands aren't clean. Don't get together with friends who are sick. And if you are sick, stay home until you are better.

Talking to friends, playing a game, or throwing a ball for your dog—your respiratory system helps you do it all. Take good care of it, and it will keep you going!

GLOSSARY

cancer (KAN-sr): A disease in which certain cells grow out of control and spread to other parts of the body

carbon dioxide (CAR-bun dye-OX-ide): A waste gas breathed out by people and animals

cycle (SAHY-kuhl): Events or occurrences that happen again and again in the same order

filter (FIL-tr): Something that makes air or water clean and pure by removing any impurities

immune (ih-MYOON): Protected from disease

membrane (MEM-brayn): A thin layer of tissue that covers part of an animal or a plant

mucus (MYOO-kuhs): A slimy substance that protects the inside of the nose and other parts of the body

organs (OR-gunz): The parts of people, animals, and plants that have special purposes

oxygen (AHK-suh-jn): A colorless, odorless gas, found mainly in air, that all plants and animals need to live

pollen (PAW-luhn): A fine yellow powder formed in flowers

structure (STRUHK-chr): Something made of many parts arranged together

tissue (TIH-shoo): The material that forms parts of animals and plants

vibrate (VAI-brayt): To move rapidly back and forth

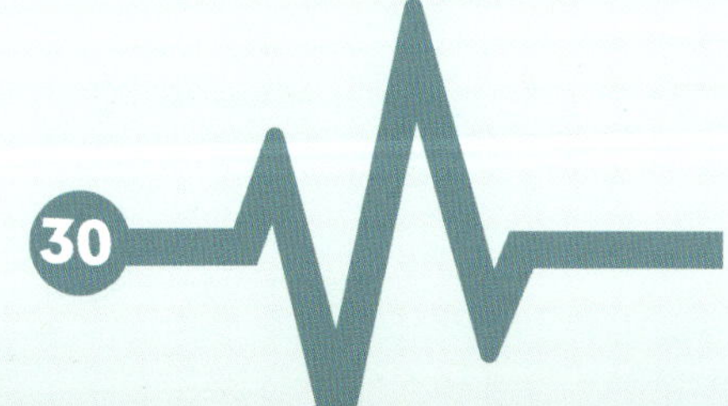

INDEX

COMPREHENSION QUESTIONS

1. What are sometimes called the building blocks of life?

a. cells
b. lungs
c. air sacs

2. The respiratory system brings in the _____ cells need to live.

a. nutrients
b. carbon dioxide
c. oxygen

3. Breathing in starts with the _____.

a. diaphragm
b. air sacs
c. vocal folds

4. True or False: Cells make a waste gas called carbon dioxide as they work.

5. True or False: Being active weakens your lungs.

Answers: 1. A, 2. C, 3. A, 4. True, 5. False

ABOUT THE AUTHOR

Tracy Vonder Brink loves to learn about science and nature. She has written many nonfiction books for kids and is a contributing editor for three children's science magazines. Tracy lives in Cincinnati, Ohio with her husband, two daughters, and two rescue dogs.

Written by: Tracy Vonder Brink
Designed by: Kathy Walsh
Series Development: James Earley
Proofreader: Janine Deschenes
Educational Consultant: Marie Lemke M.Ed.

Photographs: Shutterstock; Cover: goldnetz, Magic3D, graphicgeoff; p 3, 30 goldnetz; p 5, 10, 11, 13, 14, 15, 17, 19, 23, 28 Modvector; p 4, 6, 8, 10, 12, 14, 16, 18, 20, 22, 24, 26, 28, 30 RedlineVector; p 4, 6, 8, 10, 12, 14, 16, 18, 21, 22, 27, 28, 32 Hluboki Dzianis; p 4 crystal light, Prostock-studio; p 6 Michal Sanca; p 7 MDGRPHCS; p 8 mimagephotography; p 9 Antonio Guillem; p 10 YoloStock; p 11 Refluo; p 12 logika600; p 13 VectorMine; p 14 logika600; p 15 BlueRingMedia; p 16 Antonov Maxim; p 17 first vector trend; p 18 Olga Bolbot; p 19 Sergiy Bykhunenko; p 20 Elizaveta Galitckaia; p 21 alyaromalya; p 22 Diego Cervo; p 23 Nestor Rizhniak; p 24 Syda Productions; p 25 Krakenimages.com; p 26 Valadzionak Volha; p 27 Dragon Images; p 28 Krasula; p 29 Gorodenkoff

Crabtree Publishing

crabtreebooks.com 800-387-7650

Printed in the U.S.A./012023/CG20220815

Published in Canada Crabtree Publishing
616 Welland Ave.
St. Catharines, Ontario
L2M 5V6

Published in the United States Crabtree Publishing
347 Fifth Ave
Suite 1402-145
New York, NY 10016

Library and Archives Canada Cataloguing in Publication
Available at Library and Archives Canada

Library of Congress Cataloging-in-Publication Data
Available at the Library of Congress

Hardcover: 978-10398-0020-5
Paperback: 978-10398-0079-3
Ebook (pdf): 978-10398-0198-1
Epub: 978-10398-0138-7